YOUR KNOWLEDGE HAS VALUE

AF294621

- We will publish your bachelor's and
  master's thesis, essays and papers

- Your own eBook and book -
  sold worldwide in all relevant shops

- Earn money with each sale

Upload your text at www.GRIN.com
and publish for free

GRIN

Bibliographic information published by the German National Library:

The German National Library lists this publication in the National Bibliography; detailed bibliographic data are available on the Internet at http://dnb.dnb.de .

Imprint:

Copyright © 2016 GRIN Verlag, Open Publishing GmbH
Print and binding: Books on Demand GmbH, Norderstedt Germany
ISBN: 978-3-668-15210-6

This book at GRIN:

http://www.grin.com/en/e-book/315240/methods-at-work-in-engineering-the-weighted-matrix-pugh-matrix-and-qfd

Anna Lena Bischoff

Methods at Work in Engineering. The weighted matrix, Pugh Matrix and QFD method for decision making in product development

GRIN Publishing

Contents

1. Introduction

Only suitable methods can help to move forward in the process of product development. Therefore it is important to choose a method that fits the product, current stage of development process and the team.

The project is currently in the evaluation phase in accordance to the value methodology. One of the major purposes of this stage is to come up with a short list of projects with the highest potential. Hence, a selection process needs to be executed to produce a focused concept. The focus concept will then be used in the next stages to be developed further.

The weighted matrix is a method that is used at an early stage in product development process to select a concept that looks the most promising. This method helps to increase customer value and at the same time make a selection that is objective.

Therefore the weighted matrix was chosen in this project to help make a selection among the choice of concepts. Often it is difficult to come to a consensus among group members as everyone is biased by their profession. Everyone has a subjective way of ranking the concepts according to their experiences and knowledge. Such a ranking can lead to dissatisfaction and conflicts in the team as well as choosing a concept that has not the most potential from a customer perspective. The weighted matrix makes the decision process on the one hand a group experience where everyone is involved and gets the acknowledgement of their knowledge and on the other hand more objective.

For detailed information about the discussed company and design brief refer to "Design Brief of the case company, including company description 5.1".

2. Engineering Methods

In this chapter the value methodology is described as basic framework for this part of the project. It serves as framework with further detailing in several product development methods. Suitable tools for the project are identified and critically evaluated regarding their applicability to the project

2.1. The value methodology

The value methodology should improve the value of a product. This improvement can be either achieved by increasing the function of the product or reducing the cost. Hence, the concentration is more on the functional, value creating attributes of the to-be-developed product. The methodology proves especially worthwhile in small, multidisciplinary groups and fuzzy environments where not all parameters are set (Park, 1999). The functional analysis is the foundation of the value methodology (SAVE International, 2007). A vital part of the value methodology is the job plan, a schedule that leads through the different phases of product development with specific activities. It consists of the Information Phase, Functional Analysis Phase, Creative Phase, Evaluation Phase, Development Phase and Presentation Phase.

2.2. Theoretical framework of the weighted matrix – Pugh decision matrix

Within the value methodology a variety of specific methods can be applied for each phase. However these methods should be in compliance with the value methodology framework. This means that the methods used should look at value creation for the customer. The weighted matrix is designed to make the decision process more objective and the selection of a concept among competing concepts more respected.

The Weighted Matrix method can be applied when several design concepts have been developed and a decision enabler is needed to select promising ideas (Martin & Hanington, 2012). The Pugh Decision Matrix is a specification of a weighted matrix method. It is also called a variety of names including Pugh method, Pugh analysis, decision matrix method, decision matrix, decision grid, selection grid, selection matrix, problem matrix, problem selection matrix, problem selection grid, solution matrix, criteria rating form, criteria-based matrix, and opportunity analysis (Tague, 2004). The Pugh Decision Matrix is especially used when a decision among concepts is needed, but the concepts are not yet specified in detail and are "fuzzy" (Herrmann, 2015).

A baseline is set to which the competing concepts are benchmarked (better/positive, worse/negative, the same/zero). The matrix is made by having a list of criteria in the first column, and several options in the head column (Kubiak & Benbow, 2009). Then a criteria list with concept attributes that are important to the products design are selected. Each criterion is measured for each option in comparison to the baseline. Finally the concept with the highest ranking of Better/positive grading outcompetes the other concepts. A typical Pugh Decision Matrix is shown in the following table (Tague, 2004).

Criteria		Baseline	Concept 1	Concept 2	Concept 3
Easy handling		0	+	-	+ +
Security		0	0	- -	-
Cost		0	-	+ +	0
Σ+			1	2	2
Σ-			1	3	1

Figure 1: example of a Pugh Matrix

The Pugh matrix is used when a company has several alternative processes to the one it's using, and it wants to know if any of these is better or not (Herrmann, 2015)

It is also used when only one solution is possible, only one product can be brought to market, has only sufficient financing for one solution or where the optimal alternative is required, at the same time multiple criteria need to be considered (Tague, 2004)

It can also be used where there are many alternatives, none of which are quite suitable. The Pugh matrix can be used to choose the best aspects of the various concepts to produce a hybrid, which hopefully will be better than the alternatives used initially.

2.3. Application and evaluation of the weighted matrix – Pugh decision matrix

The Pugh decision matrix concentrates on qualitative decisions (Samuel & Weir, 2005). Therefore it interlinks well with the value methodology where emphasis is put on the added value for the customer. Furthermore it is often used at early stages of decision making (Farag, 2014), which applies to the current situation of the project team.

When applying Pugh decision matrix, the first thing that needs to be done is to have a clear understanding of the existing concepts. The team members have generated three concepts: concept 1, concept 2 and concept 3. The main materials of these concepts are fiber board, which is the only similarity of these concepts. The difference is the constructional structure.

Concept 1 uses wood frames to support the coffin and make it more stable.

Concept 2 consists of fiber board only. It has some ingenious structures, like a double layer to support the coffin and make it stable.

Concept 3 is the combination of concept 1 and 2, basing the double-layers structure, adding wood frames to provide stronger capacity, at the same time, it may cost more.

To make a Pugh decision matrix, after having the concepts, the next step is to choose the baseline. For this project, the traditional coffin was used as the baseline. It has traditional shapes as most of western coffin in the market used the same shape; also it is totally consist of solid woods, which can provide tough strength.

Criteria	Baseline	Concept 1	Concept 2	Concept 3
Environment impact	0	+	++	+
Cremation	0	-	--	-
Easy carry	0	0	+	0
Costs	0	+	++	+
Innovation	0	++	++	++
Material Strength	0	-	--	-
Sustainability	0	+	+	+
Qualities of Material	0	-	--	-
Affordable	0	+	++	+
Σ +	0	6	10	6
Σ −	0	3	6	3

Figure 2: Pugh decision matrix evaluated by the team on their concepts

From the above table, it is obvious that the alternative concepts have different marks basing the same baseline on each criterion. For instance, concept #1 and #3, which are consist of both woods and cardboard would have a better environment impact than the baseline coffin, because the degradations of cardboard would be easier than solid woods. As the same, the concept 2, which is totally consist of cardboard, would performs better than concept 1 and 3, thus in the above table, the concept 1 and 3 have a '+' and concept 2 have '++'.

The team gains from the Pugh decision matrix is that it shows a valuable way to choose a suitable concept for further development from several generated concepts, it provide a convincingly method to persuade every team member and avoid arguments. For this project, it is obvious that concept 2 performs better than the other two.

However it is criticized that the Pugh decision matrix is a quick overview and does not include much detail when forming the decision as it does not weight the importance of each criteria (Herrmann, 2015). This needs to be considered when using the matrix being aware that it is limited and certain criteria might be more important than others to the end-consumer. Examples can be that a customer is more interested in the environmental impact of the product as of the innovative qualities.

3. The methodology process and phases

After having set the methodological framework in chapter two, the projects engineering process will be described. Starting with a quick review by using the time schedule for the project, followed by the detailed process description and result documentation.

3.1. Project schedule

The project schedule was set up during Module 2, design according to the product development phases of Martin and Hanington (2012). Figure 1 describes were the project team is in its development process. Concepts need to be evaluated and tested to complete the Module 3, engineering successfully and move into the next module, business. The time set and weeks changed during the process of the design module. However, the tasks that the project team set itself remained the same. Currently the project schedule has two weeks delay. The part of "Launch and Monitor" will probably not be completed in this module.

Part of Module 3 and 4: engineering and business

	Planning, Scoping and Definition	Exploration, Synthesis and Design Implications	Concept Generation and early Prototype iteration	Evaluation and Refinement	Launch and Monitor
Principal activities	Interview with AB Understand the current state of AB Validate requirements from the brief Evaluate interview Questionnaire Set up project schedule	Analyze the major competitors Define substitutes Evaluation of Home Office Supply market SWOT analysis Analyze materials used Reset scope into new direction	Start sketching and redo Select design idea concepts Develop detailed design Develop flat pack solution Do a weighted matrix analysis Generate and prioritize recommendations with AB	Implement the feedback of AB in the design Agree approach and refine requirements Start Prototyping, experimenting and testing Customer requirements and variations validations Finalize design and gain approval	Develop implementation plan Decision point of sale Tracking of sales , customers satisfaction
Work results	Interview notes Time schedule Clear design goal	SWOT of AB Two design concepts Market analyses	Two decided concepts and target group Weighted matrix with concept decision	Prototype Approved testing Approved design	Product ready to launch Product recommendation
	Week 41	Week 42 to 43	Week 44 to 47	Week 47 to 49	Week 50 to 53

Figure 3: Time schedule for project The case company

3.2.   The methodology process

After analysis the three different concepts a weighted matrix was applied in order to select the best concept, which would add the most value to the customer. This method helped to give a better understanding for the different concepts. In the following paragraphs will be described how the rating and selection process was conducted.

Choosing the baseline
The baseline was selected by taking the most common type of casket: a wood casket.

Choosing the criteria list
When conducting a Pugh analysis, criteria are chosen that will be measured against the baseline for each concept. The following criteria were selected in accordance with value adding functions for the customer:

Environment Impact
This criterion evaluates how the product influences the environment throughout the product life cycle. There should be no contamination from the coffin or corpse when released into the soil. It is an integral part of the product life cycle besides ecologic material sourcing and production process.

Easy carriage/handling
This attribute considers how difficult or easy it is to lift the product as cemetery and funeral home personnel will have to carry the coffin. The measure is based on the total weight of each concept caused by the different materials build in each concept.

Cost
Costs are mainly based on the cost of materials for each concept. This criterion does not reflect the material consumption. This attribute is important to see the liability of the concept in terms of profitability and investment.

Innovation
This criterion represents how the innovation of the product life cycle will be connected with the environmental impact and the materials decomposition into the soil. All three concept are innovative in regards to the product's life cycle. Use-time and decomposition time can be reduced due to materials used as they will consume less time to decompose. Hence grave plots can be reused quicker. Another major innovative factor is the improved environmental impact of the concepts.

Material strength
Here the materials are assess by their strength to support the corpse without any risk of breaking. It is necessary as the corpse should be carried safely from car to chapel and

chapel to grave plot. Therefore, the materials chosen has to be stiff enough, but absorb the vibrations of transportations as well.

Sustainability
This attribute covers a wide set of aspects, where the product's life cycle has impacts on people and environment. Sustainability describes more the effect on how long a product will last and be in use than the sole "green-ness" of it.

Quantity of materials consumed
The quantity of materials that each concept consumes varies. Therefore this criteria asses the resource expenditure of each concept as the amount consumed is closely linked to costs and sustainability.

Affordability
The products should be more affordable to customers without compromising on the aesthetics and function. This will offer the product an advantage to competitors.

Cremation
Coffins are not only buried in the ground, but also cremated. The environmental impact of burning certain materials such as wood versus fiber board is considered in this criterion.

Executing the evaluation
Figure 2 describes how the group members developed the matrix and ranked the concepts. It is then described in detail why which concept was graded at what level and why.

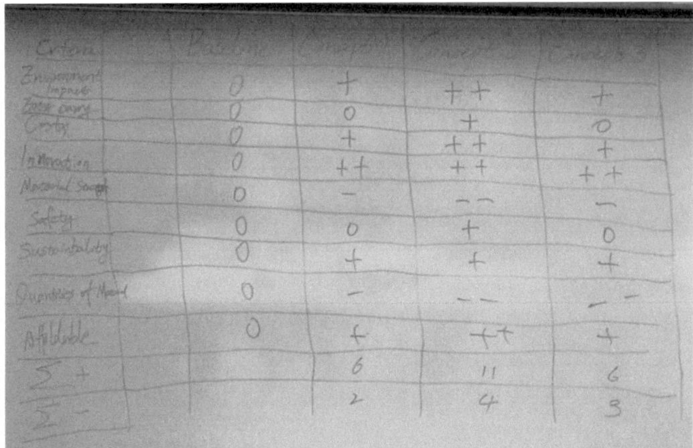

Figure 4: Picture of the development of the Weighted Matrix method

Environmental impacts
Of the three concepts, the second has the lowest environmental impact because it requires one material: the fiberboard which is recycled and recyclable. According to Eco-cerc (ECO-CERC, 2015), a French company distributing cardboard coffins, the effect of cardboard coffins could save a high quantity of resources as described in figure 3.

Per day	Per year
86,4 km² forest	31 536 km² forest
32 640 healthy trees	11 913 600 healthy trees
18 240 m3 water	6 657 000 m3 water
864 000 liter of fuel	315 360 000 liter of fuel

Figure 5: Resources saved when using cardboard instead of wood for coffins

Cremation
The incineration of a cardboard coffin takes half the time of a wooden coffin which rejects toxic micro particles dangerous for the health, the cremation of cardboard is not harmful to health, but it has a disadvantage, in fact it consumes more gas than wood, gas is a nonrenewable fossil energy, exacerbating the greenhouse effect.

Easy carriage/handling
As a wood coffin the fiberboard coffin is carried thanks to handles. The weight of the body makes the coffins difficult to carry and manipulate. However, cardboard is lighter than wood so the concept 2 which is entirely made of cardboard would be easier to carry.

Cost
The cost of production for the fiberboard coffin, (concept 2) would be the cheapest of the three because it requires only one material.

Innovation
The fiberboard coffin is an ecological innovation; the concept is more and more popular. It offers lots of advantages such as low environmental impacts, affordable price, easy to personalize. By using this concept, we are trying to add a new innovative construction to make the coffin more resistant.

Material strength
Cardboard is a light material, when it's not treated properly it dread humidity and is fragile, it don't have a lot of wear. The strength of the coffin will depend on the

innovative construction in progress during the engineering module. For concept 1 and 3, an internal wood structure would make the coffin stronger.

Sustainability
Once more the concept 2 is the best option. Indeed, a cardboard coffin is biodegradable, made of recycled materials and recyclable. The waste prevention and recycling of cardboard can reduce the use of resources and the environmental impacts related to its depletion.

Quantity of materials consumed
All of the three propositions requires an amount of cardboard in order to make the coffin strong and waterproof. But concept 1 and 3 will require even more material because they are reinforced by a wood structure

Affordability
The price of a cardboard coffin varies from 100 EUR to 600 EUR instead of 800 EUR to 3000 EUR for wood coffin. The concept 2 would have the most reasonable price

By confronting the three concepts in this method, it allowed the group to understand that the second concept is the most advantageous on many levels: a coffin entirely made in fiberboard is the more profitable for the majority of the analyzed criteria.

4. Conclusion

The Pugh analysis is a helpful method in a product development process when several concepts exist and the team is disagreeing which concept should be further pursued. It can be used early on in the process, helps to add customer value in the decision and make the selection more objective. The objectiveness is important as each team member is biased by their profession's knowledge and preferences. The analysis also shows where in the competing concepts there are the major strengths and weaknesses.

In the evaluation phase of the value methodology it is a very useful tool. It could be further improved if considering the criteria by which the concepts are measured would be weighted by their importance. The importance could be measured by customer perception or expectation. To weight the criteria would also help to transfer the product from engineering phase into business were the customer needs to be perused to buy this product. Persuasion is easier when customer requirements are met in a product.

5. Appendix

5.1. Design Brief of the case company, including company description

This is just a suggestion of questions formulations that you can work with. We can together talk about if your background our earlier experience would be better to focus on some of this questions and also what is reasonable for your timeline. It is important that you all feel inspired.

Objective of the project

What does the project requires from the business perspective?

* How should we marketing and sale the new products. Is our idea of sell it on the web the best way to go? Create a business and marketing plan. How is our target?

* Make a suggestion/ a plan of how we should work with advertising, also a budget for this.

* Calculate how much the production cost is, row material and what should we sale it fore?

* Brand, how can we make our brand stronger?

* How can we make the production and product more cost-effective?

* How are our competitors? What can we do to stand out from them? What do we offer that they do not do, and how can we show this for our costumer?

What does the project requires from the engineering perspective?

* Construction of the box, dare to question the flat packing solution on the market today. How can we make it so easy for both in the production and also for the costumer?

* Production, can we make the production of the flat packing line more effective? Technique, machines, details.

* Sustainability, how can we work with sustainability in a better and more effective way? Also how can the customer easy separate the different

* Explore what sizes there boxes should be.

What does the project requires from the design perspective?

* Material, fiber board paper, leather and metal details.

* How is the user for this product? And how will he/she use it? What are our users' needs and how do they perceive the product. How are the different stakeholders for the project? Try to involve them in to the project.

* Sizes, investigate witch sizes our user need. How many part should there be in the collection? Shape, color, prints etc.

* A construction that holds the boxes together, how can you make it in to a detail? Work with construction, function and shape..

* If a manual is necessary for the costumer to put the boxes together, make a manual that fit the graphic profile and is easy to understand (illustrated).

* Sustainability, how can sustainability be a part of this project? Material, construction, etc.

Our company's wishes, feelings and expectations

* Our wishes are that your group creates something that fits both the private and the public sector.

* It is important that it is a clear line between all different elements. Try to involved and be inspired of our key word is: History, craftsmanship and Sweden.

Timing and approvals

This is given by us according to the course schedule

5.2. Table of Figures

5.3. Bibliography

Bryman, A., & Bell, E. (2011). *Business Research Methods*. OUP Oxford.

ECO-CERC. (2015). *Some numbers*. Retrieved 11 2015, 20, from Ecocerc: http://www.eco-cerc.fr/ecocerc/respect-de-l-environnement/

Farag, M. M. (2014). *Materials and Process Selection for Engineering Design*. New York: Taylor & Francis Group.

Herrmann, J. W. (2015). *Engineering Decision Making and Risk Management*. New Jersey: John Wiley & Sons, Inc.

Kubiak, T. M., & Benbow, D. W. (2009). *The Certified Six Sigma Black Belt Handbook*. Milwaukee: Quality Press.

Liou, F. W. (2007). *Rapid Prototyping and Engineering Applications: A Toolbox for Prototype Development*. New York: Taylor & Francis Group.

Martin, B., & Hanington, B. (2012). *Universal Methods of Design: 100 Ways to Research Complex Problems, Develop Innovative Ideas, and Design Effective Solutions*. Rockport Publishers.

Park, R. (1999). *Value Engineering: A Plan for Invention*. New York: CRC Press LLC.

Samuel, A., & Weir, J. (2005). *Introduction to Engineering Design*. Oxford: Elsevier Butterworth-Heinemann.

SAVE International. (2007). *Value Methodology Standard and Body of Knowledge*. SAVE International.

Schön, D. A. (2008). *The Reflective Practitioner: How Professionals Think in Action*. Basic Books.

Tague, N. R. (2004). The Quality Toolbox.

Ulrich, K. T., & Eppinger, S. D. (2012). *Product Design and Development*. New York: McGraw-Hill.

YOUR KNOWLEDGE HAS VALUE

- We will publish your bachelor's and master's thesis, essays and papers

- Your own eBook and book - sold worldwide in all relevant shops

- Earn money with each sale

Upload your text at www.GRIN.com and publish for free